Impact on Earth

The Impact of
ENERGY

Nancy Dickmann

Crabtree Publishing Company
www.crabtreebooks.com

CRABTREE
PUBLISHING COMPANY
WWW.CRABTREEBOOKS.COM

Author: Nancy Dickmann

Editorial director: Kathy Middleton

Editor: Ellen Rodger

Picture Manager: Sophie Mortimer

Design Manager: Keith Davis

Children's Publisher: Anne O'Daly

Production coordinator and prepress: Ken Wright

Print coordinator: Katherine Berti

Photo credits
(t=top, b= bottom, l=left, r=right, c=center)

Front Cover: Shutterstock / connel, br; All other images from Shutterstock

Interior: Alamy: Orfan Ellingvag 11, Adam Murphy 23, US Coast Guard Photo 15; energy-floors.com: 5; Getty Images: AFP/ Ali Arif 7; iStock: fotografixx 27, gyn9038 26, Mimadeo 17, poweroforever 19, Michael Utech 16, william87 24; Shutterstock: Africa Studio 29, Bannafarai-Stock 14, BlurryMe 9, Hywit Dimyadi 6, Mr. JK 4, Monkey Business Images 25, Kamil Petran 13, Evgeny Petrushin 28, PhotographyByMK 22, CL Shebley 20, Hiroshi Teshigawara 12, Evgeny V 18, Tommy Lee Walker 8, Yermolov 10.

Library and Archives Canada Cataloguing in Publication

Title: The impact of energy / Nancy Dickmann.
Names: Dickmann, Nancy, author.
Description: Series statement: Impact on Earth | Includes index.
Identifiers: Canadiana (print) 20190236752 |
  Canadiana (ebook) 20190236760 |
  ISBN 9780778774358 (hardcover) |
  ISBN 9780778774501 (softcover) |
  ISBN 9781427125149 (HTML)
Subjects: LCSH: Power resources—Juvenile literature. |
  LCSH: Power resources—Environmental aspects—Juvenile
  literature. | LCSH: Renewable energy sources—
  Juvenile literature.
Classification: LCC TJ163.23 .D53 2020 | DDC j333.79—dc23

Library of Congress Cataloging-in-Publication Data

Names: Dickmann, Nancy, author.
Title: The impact of energy / Nancy Dickmann.
Description: New York, New York : Crabtree Publishing Company,
  [2020] | Series: Impact on Earth | Includes bibliographical
  references and index.
Identifiers: LCCN 2019053357 (print) | LCCN 2019053358 (ebook)
  ISBN 9780778774358 (hardcover) |
  ISBN 9780778774501 (paperback) |
  ISBN 9781427125149 (ebook)
Subjects: LCSH: Power resources--Juvenile literature. | Energy
  consumption--Juvenile literature. | Renewable energy sources-
  -Juvenile literature. | Human ecology--Juvenile literature. |
  Climatic changes--Effect of human beings on--Juvenile literature.
Classification: LCC TJ163.23 .D53 2020 (print) |
  LCC TJ163.23 (ebook) | DDC 333.79--dc23
LC record available at https://lccn.loc.gov/2019053357
LC ebook record available at https://lccn.loc.gov/2019053358

Published in 2020 by Crabtree Publishing Company

Printed in the U.S.A./022020/CG20200102

Published in Canada
Crabtree Publishing
616 Welland Avenue
St. Catharines, ON
L2M 5V6

Published in the United States
Crabtree Publishing
PMB 59051
350 Fifth Ave, 59th Floor
New York, NY 10118

Contents

Using Energy

Energy is everywhere! Light and heat are both forms of energy. Energy from the Sun travels to Earth in these forms.

We use energy in many different ways. You use your own body's energy when you ride your bike. A lightbulb uses energy to brighten a room. Cars use energy to drive down the road.

Cities need a huge amount of energy to heat and light buildings and move people from place to place.

Dance Power

When you dance, you turn your body's energy into movement energy. We usually don't try to collect this movement energy for our use, but one company has invented a dance floor that captures the energy. It converts movement on the dance floor into **electricity**. The electricity can power the lights and sound system for the room.

In 1 day, the Sun sends out more energy than the world uses in a year

93 million miles (150 million km)
The distance energy travels from the Sun to reach Earth

A World of Machines

More than a hundred years ago, people relied mainly on their own energy or the energy of animals to travel or do work. Some places began to use mills to harness the energy of wind or flowing water. Today, we are surrounded by machines that run on energy and make our lives easier.

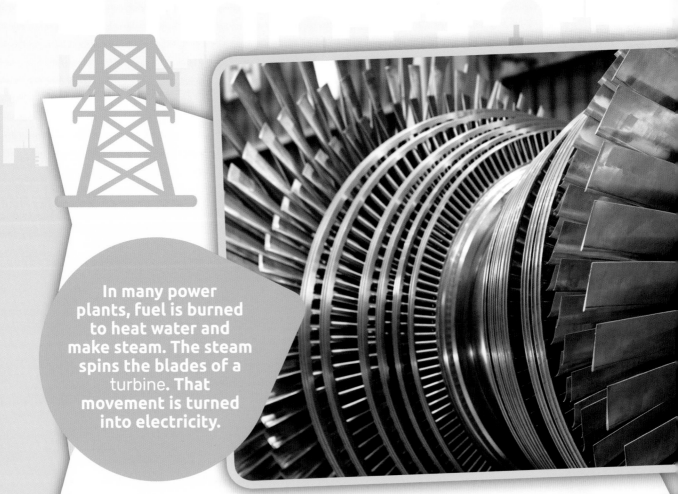

In many power plants, fuel is burned to heat water and make steam. The steam spins the blades of a turbine. **That movement is turned into electricity.**

Electricity

A lot of the energy we use is in the form of **fuels** such as gasoline, which is burned in cars and trucks to power the motor. But nearly 20 percent of the total energy we use is in the form of electricity. Most electricity is produced in power plants. We can **generate** electricity by burning fuels such as coal. We can also convert sunlight or the movement of wind or water into electricity.

Easy Electricity

Most homes in **industrialized** countries have a wire that connects them to an **electricity grid**. This gives us a constant supply of power that we can use whenever we need. Whether it's turning on a light, microwaving a snack, powering a guitar amp, or charging an electric car, we use electricity every day.

37% Percentage of total electricity produced that is used in homes in U.S.A.

EMERGENCY · EMERGENCY·

Blackout!

When there is a problem with the electricity supply, it can affect a lot of people. In India on July 30, 2012, a series of problems led to a failure of the electricity grid. Over 300 million people lost power. The problem was fixed, but the next day another blackout left 600 million people without power.

Why Worry?

The world's population is growing and we need more energy. We are also becoming more dependent on devices that use energy.

Traffic lights, machines in factories, computers, and washing machines didn't exist hundreds of years ago. Today, in many countries, we rely on them, and we need more energy to run them. More power plants are being built to keep up with growing demand.

The files that make up the Internet are stored on collections of computers called server farms. These farms use electricity to run the servers and keep them cool.

Using Less

We can all help ease the growing demand for energy. Driving less and using bikes or public transportation will reduce the amount of fuel used. You can also cut down on energy use at home. Wear a sweater, turn down the thermostat, or switch to low-energy lightbulbs.

1970-2019 The world's population doubled over this period

2000-2019 The world's electricity use has nearly doubled

Running Out

We use coal, oil, and natural gas to meet most of our energy needs. These are called **fossil fuels**. They are extracted, or taken out of the ground, by drilling. Fossil fuels are not **renewable**. That means they cannot be replaced. Once we have used them up, we will have to find new sources for power.

Harming the Planet

Burning fossil fuels—either in vehicle engines or in power plants—releases waste gases. One of these gases is **carbon dioxide**. Too much carbon dioxide in the **atmosphere** makes the planet warm up. Burning fossil fuels also produces tiny pieces of matter, called particulates. Breathing waste gases and particulates can damage your health.

We mine coal to use its energy, but just getting it out of the ground uses up a lot of energy. It also destroys important habitats for plants and animals.

Capturing Carbon

Scientists around the world are working on ways to remove carbon dioxide from the air. One company is taking this a step further by using the carbon dioxide they capture to make fuels that can be used in cars, trucks, and airplanes.

About 66% Percentage of electricity produced by burning coal, oil, or gas

1,300 Number of Olympic-sized swimming pools that could be filled by the amount of oil used daily in U.S.A.

Greenhouse Gases

When too much carbon dioxide builds up in the air, it forms a layer in the atmosphere. This traps the Sun's rays, just like the glass in a greenhouse does. Because the rays can't bounce back into space, Earth's temperature gradually increases, affecting climate around the world.

CO_2

Fossil Fuels

There are three main types of fossil fuels that are used to generate electricity: coal, oil, and natural gas.

All three are formed from the remains of living things that died millions of years ago. The remains were covered over with layers of earth. As it decayed, or broke down, over time, heat and pressure turned the material into fuel.

Coal is a solid material and looks like rock. Crude oil, shown here, comes out of the ground as a liquid. Natural gas is a colorless gas.

Smokestack Scrubbers

Fossil fuel power plants burn carbon. This produces carbon dioxide, which is a greenhouse gas. Power plants are able to reduce the amount that goes into the air. One coal-fired power plant in Texas has a "scrubber" on one of its smokestacks that removes about 90 percent of the carbon dioxide before it can escape.

Less than 50%

Percentage of the energy in fossil fuels burned that is actually converted to electricity in most power plants

Power Plants

A power plant turns one type of energy into another. A small amount of coal, oil, or gas contains a lot of energy. When it burns, it releases energy as heat, which boils water to produce steam. The steam spins a turbine. This movement energy is converted to electricity.

Which Is Dirtiest?

In the United States, about 63 percent of all electricity is produced by burning fossil fuels. Not all fossil fuels are the same. Coal is by far the biggest polluter. It produces the most carbon dioxide when it is burned. It also releases other chemicals harmful to our health. Coal is easier to extract than oil or gas, so a lot of it gets used.

For a long time, coal has been the main fuel used around the world to generate electricity. Today, natural gas is catching up quickly.

Cleaner Gas?

Natural gas is cleaner to burn than coal, but it still produces carbon dioxide. It sometimes leaks when it is extracted or sent through a pipeline to another destination. This releases other gases that contribute to **climate change**. Oil is widely used for heating and as a fuel for vehicles, but it is only rarely used in power plants.

27% in U.S.A. and 9% in Canada Percentage of country's electricity produced by burning coal
38% Percentage of electricity produced worldwide by burning coal

EMERGENCY · EMERGENCY ·

Oil Disasters

Creating waste gases is not the only problem with fossil fuels. The Deepwater Horizon was an oil drilling platform in the Gulf of Mexico. In 2010, an explosion killed 11 workers and caused oil to spill into the water for almost three months. The oil damaged huge areas of coastline, killing plants and animals.

Renewable Energy

There are other ways to produce electricity that have less impact on the planet.

Renewable energy means a source of energy that can be replaced and will never run out, such as sunlight. Most types of renewable energy also produce little or no **pollution**. They don't release much carbon dioxide into the air, which is why they are described as being "clean" energy.

Nuclear power is often called "clean" because it releases little carbon dioxide. However, it uses uranium, a type of fuel that is not renewable.

WHAT CAN I DO?

Renewable or Not?

Depending on where you live, you may be able to choose whether your electricity comes from renewable sources. It may cost slightly more, or it may be the same price as traditional electricity. Ask an adult to find out where your electricity comes from, and whether switching to a renewable energy is an option.

24% Percentage in 2017 of world's electricity made from renewable energy

500 Number of homes one wind turbine can generate energy for

Wind Power

Wind turbines turn the movement energy of wind into electricity. Wind spins the blades of a huge turbine, which then spins a **generator** that creates electricity. Wind turbines can be built on land or out at sea, where winds are often stronger. However, they only work when the wind is blowing.

Electricity from Water

The movement energy of flowing water can also be turned into electricity. In a hydroelectric plant, a dam blocks the flow of a river. It forces the water through turbines, making them spin. As the turbines spin, they generate electricity. Dams can be small, powering just a dozen homes or so, or they can be big enough to light a city.

Dams are expensive to build, but once they are up and running, the cost of producing electricity is fairly low.

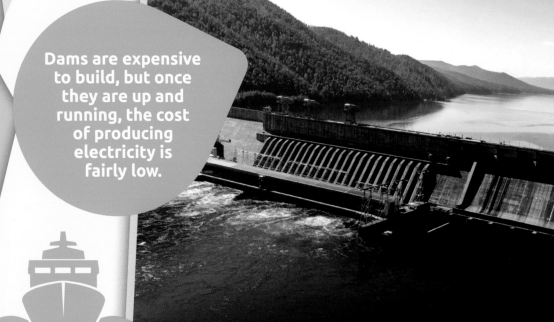

Dam Damage

Building a dam creates a huge reservoir, or pool, of water for nearby towns and cities. China needed a reservoir so huge it required the flooding of more than 1,500 towns and villages. At least 1.3 million people had to leave their homes for the Three Gorges Dam to be built.

About 16% Percentage of world's energy that comes from hydroelectric power

57% Percentage of Canada's electricity from hydroelectric power

Ocean Energy?

A few power plants rely on tides, the daily rising and falling of the oceans. They are often built where a river meets the sea. Engineers are also working on efficient ways to use the power of waves. The up-and-down movement of the swells could be turned into electricity.

Electricity from Heat and Light

Some solar panels capture the energy in sunlight. They have special cells that can turn light into electricity.

These panels are often set up in huge groups called solar farms. The panels can be expensive, but once installed, solar power is cheap to produce. On the downside, the panels don't work at night, and the electricity they produce is hard to store.

A house can often meet its electricity needs by putting solar panels on the roof. Smaller solar panels can charge phones and electronic devices.

Steam from Underground

Deep underground, it is very hot. We can turn Earth's heat into electricity. At a geothermal power plant, deep wells are drilled into the ground. Hot water and steam are piped up to the surface to spin a turbine. When the steam cools back into water, it is returned to the ground.

2.5 million Number of solar panels at the Kamuthi solar power plant in India

750,000 Number of people who get their electricity from Kamuthi

TECHNOLOGY SOLUTIONS

Heat Pumps

Geothermal heat is also useful for heating homes. A ground source heat pump uses pipes buried below a home's yard to bring heat from the ground into the house. In summer, it cools the house by taking heat back into the ground. Some systems use shallow pipes, while others drill much deeper.

Biofuels

We often burn plants as fuel, such as the wood for a campfire. Plants can also be turned into fuel for vehicles. These fuels are called **biofuels** because they come from living things. Corn and sugarcane are two of the main crops used to make a biofuel called ethanol. Because new plants grow to replace the ones we use, the fuel is renewable.

To produce biofuels, farmers use their land to grow crops for fuel instead of food. Less food grown could mean food prices will go up.

Recycling Food Waste

Don't throw away your food waste! In many places, you can recycle it. Banana skins, coffee grounds, eggshells, and stale bread can all be recycled. This waste is often turned into fertilizer. It can also be broken down to produce a biogas that is burned to generate electricity.

85% Percentage of the world's ethanol produced by U.S.A. and Brazil

Over 98% Percentage of gasoline sold in U.S.A. that contains some ethanol

Cleaner Fuels

Biodiesel is another type of biofuel. It is usually made from vegetable oils, such as soybean oils. It can even be made from **recycled** fats, such as cooking oil from restaurants. Both types of fuel release carbon dioxide when they burn, but they release less than fossil fuels.

Future Developments

Our planet faces great challenges from climate change. We can help by releasing less carbon dioxide and other greenhouse gases.

Transportation and producing electricity account for about 40 percent of greenhouse gas emissions. The remainder comes mostly from farming and factories. We need to find cleaner ways to live, work, and eat.

Everything that you eat or buy had to be produced and transported. All of these actions use energy. Buying less will reduce the energy used.

WHAT CAN I DO?

Cutting Down

There are a lot of things you can do to reduce your energy use. Simple changes like turning lights off in an empty room can help a little. Bigger changes such as eating less meat, because raising animals uses a lot of energy, or taking public transportation instead of using cars, will have a bigger effect.

About 90% Percentage increase of carbon dioxide emissions since 1970

About 42% Percentage of the world's carbon dioxide emissions produced by U.S.A. and China

Using Less, Using Smarter

Using less energy overall would be a first step. This could mean traveling less or buying fewer things. Engineers are also working on making vehicles, appliances, and power plants more efficient. If they are efficient, it means that they use less energy to get the same results.

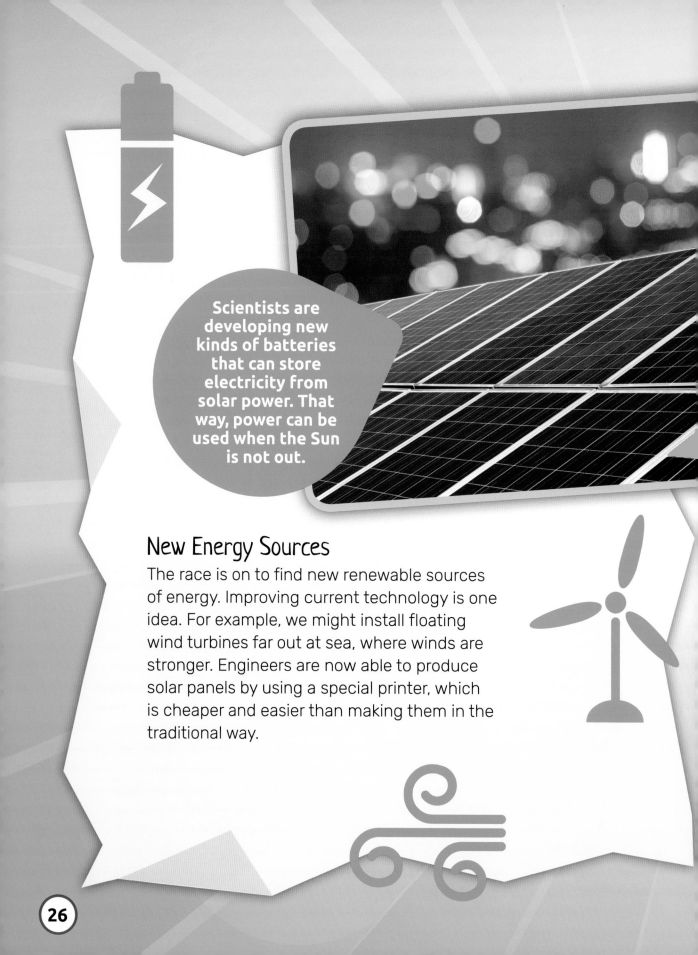

Scientists are developing new kinds of batteries that can store electricity from solar power. That way, power can be used when the Sun is not out.

New Energy Sources

The race is on to find new renewable sources of energy. Improving current technology is one idea. For example, we might install floating wind turbines far out at sea, where winds are stronger. Engineers are now able to produce solar panels by using a special printer, which is cheaper and easier than making them in the traditional way.

What's Ahead?

Energy in the future might look very different from the picture today. One day we might produce clean energy through a process called nuclear fusion. We might drive cars powered by biofuels made from algae. Or we might fly in planes powered by **hydrogen** fuel cells, which use hydrogen as fuel.

32% **The goal of the European Union is to produce 32% of all its energy from renewable sources by 2030**

Green Power!

Algae are plant-like living things. They produce energy from sunlight, like plants do. They take carbon dioxide out of the atmosphere as they grow. Scientists are studying some types of tiny algae that store the energy as natural oils. If we farm algae, we can harvest the oils and use them as fuel.

Your Turn!

How could you reduce the amount of energy that you use at home or in school?

Gathering Evidence

Start by taking a survey to find out how people use energy. You could interview people or make a questionnaire. Here are some questions that you might want to ask:

- How often do you take a trip in a car? What about on a bicycle or public transportation?
- How do you heat your home?
- Does your home have good insulation that keeps heat or cool air from escaping outside?
- How do you dry your clothes?
- Do you turn off lights when you leave the room?

Think about ways to ask questions that will give simple answers. They will be easier to sort and analyze.

Electric dryers use a lot of energy. Hanging clothes outside to dry saves energy and leaves them smelling fresh.

Results

Once you have your answers, analyze the results to get the bigger picture. You might be able to show the results in a graph or chart. What do they tell you about the way that people use energy?

Finding Solutions

Can you see any ways that energy is being wasted? Can you suggest any lifestyle changes that might mean people drive less? Are there steps that can be taken to make a building more energy efficient? Installing solar panels or pumps to collect underground heat will reduce the carbon dioxide you put out, but that can be expensive. Are there cheaper ways people could make a difference?

Glossary

atmosphere The layers of gases that surround Earth

biofuels Fuels produced from living things such as plants

carbon dioxide A gas found naturally in the atmosphere, which is also produced when living things exhale and fossil fuels are burned

climate change A change in climate patterns around the world due to the warming of Earth by greenhouse gases

electricity A form of energy that can power devices such as lights and computers

electricity grid A network of wires and power plants that delivers electricity to buildings

emissions Gases or other substances that are given off when fuel is burned

fossil fuels Fuels that are formed from living things that died millions of years ago

fuels Substances burned to produce heat or power

generate To produce electricity from another form of energy

generator A machine that changes movement energy into electrical energy

greenhouse gases Gases such as carbon dioxide that build up in the atmosphere and trap heat

hydrogen A colorless, tasteless, odorless gas

industrialized Having industry and technology

pollution The adding of harmful substances into an environment

recycled Used again in its original form or in a new form

renewable Able to be replaced and will never run out

turbine A device with blades that spins when air or water flows past it

Find Out More

Books

Dickmann, Nancy. *Using Renewable Energy (Putting the Planet First)*. Crabtree Publishing, 2018.

Dodd, Emily. *DKfindout! Energy*. DK Children, 2018.

Eboch, M. M. *The Future of Energy: From Solar Cells to Flying Wind Farms (What the Future Holds)*. Capstone, 2020.

Sneiderman, Joshua and Erin Twamley. *Renewable Energy: Discover the Fuel of the Future with 20 Projects (Build it Yourself)*. Nomad Press, 2016.

Websites

Visit this website to find out about climate change and the greenhouse effect.
climatekids.nasa.gov/greenhouse-effect/

This website has lots of information about what energy is, how it is made, and how it is used.
www.eia.gov/energyexplained/

Find out how electricity works on this fact-filled website.
science.howstuffworks.com/electricity.htm

Index